HANDY POCKET GUIDE TO
Asian
Vegetables

Text and recipes by Wendy Hutton
Photography by Peter Mealin

GW00536755

PERIPLUS EDITIONS
Singapore • Hong Kong • Indonesia

Published by Periplus Editions with editorial offices at
61 Tai Seng Avenue, #02-12, Singapore 534167.

www.periplus.com

Copyright © 2004 Periplus Editions (HK) Ltd.
ALL RIGHTS RESERVED

ISBN: 978-0-7946-0194-2

Distributors
Indonesia
PT Java Books Indonesia
Kawasan Industri Pulogadung
Jl. Rawa Gelam IV No. 9
Jakarta 13930
Tel: (62) 21 4682-1088
Fax: (62) 21 461-0206
crm@periplus.co.id
www.periplus.com

Japan
Tuttle Publishing
Yaekari Building, 3rd Floor
5-4-12 Osaki, Shinagawa-ku
Tokyo 141-0032
Tel: (81) 3 5437-0171
Fax: (81) 3 5437-0755
sales@tuttle.co.jp
www.tuttle.co.jp

Asia Pacific
Berkeley Books Pte Ltd
61 Tai Seng Avenue #02-12
Singapore 534167
Tel: (65) 6280 1330
Fax: (65) 6280 6290
inquiries@periplus.com.sg
www.periplus.com

Printed in Malaysia   1410TW

19 18 17 16 15 14
8 7 6 5 4 3

Introduction

One of the most striking aspects of markets throughout tropical Asia is the vast array of vegetables on display: bulging beige, white, purple or pink roots, tubers and stems; round, oval or snake-shaped gourds; dozens of vivid leafy greens, often glistening with the water sprinkled on to keep them fresh; tubs of crisp beansprouts and piles of bamboo shoots; forest-fragrant mushrooms and bunches of aromatic greens; mountains of fiery chillies and even edible flowers.

Some of the vegetables enjoyed in Southeast Asia are rarely seen in markets, as they are gathered wild, plucked from the tips of trees, from alongside rivers or *klongs*, from the edges of rice fields or from the forests. Others are from kitchen gardens and are rarely grown commercially. As Asia becomes increasingly urban, the use of these wild vegetables is diminishing—as, indeed, is the prevalence of kitchen gardens—yet the markets offer more than enough to compensate for the lack of these traditional vegetables.

Many of tropical Asia's vegetables are native to the region. Countless others have been introduced from other continents and are now such an accepted part of the local diet that it's hard to imagine a time when they were not available. The most striking example is the chilli, a fiery little fruit unknown outside its native tropical America before the arrival of Columbus in the 15th century. Portuguese and Spanish colonialists carried the chilli to India and to the Philippines, from whence it spread like the proverbial wild fire.

This book serves as an introduction to some of the most common vegetables found in tropical Asian markets, and includes only those Western vegetables whose Asian counterparts are somewhat different. Most of the entries are lowland vegetables; others are cultivated in cooler highland areas or imported from China.

Local names usually differ from one country to the next; even within countries such as Indonesia, the Philippines and Thailand, the same vegetable may have different names according to the local dialect. The most accurate identification is the scientific name, although even here, botanists show a reluctance to let sleeping plants lie and sometimes reclassify a variety. This book gives the currently accepted botanical name, with older or alternative botanical names also provided where these may be useful. The most widespread local name is given for Thailand, Malaysia, Indonesia and the Philippines. The vegetables are grouped according to family, in alphabetical order. An index which includes their common and scientific names can be found at the back of the book.

Identifying vegetables is only the first step towards their enjoyment. Buying tips, storage and preparation are all discussed, as well as notes on the history, nutritional value and medicinal properties of certain vegetables. Finally, a few sample recipes from around the region are included to whet your appetite.

Shiitake Mushroom

Lentinus edodes

Botanical Family:
Agaricaceae

Thai name:
Hed horm

Malay name:
Cendawan

Indonesian name:
Jamur

Filipino name:
Kabuteng kahoy

Until relatively recently, this north Asian mushroom, also known as the black forest mushroom, was found only in its dried state in tropical Asia. Fortunately, the fresh mushroom, generally sold under its Japanese name, *shiitake*, is increasingly grown in the region and is highly prized for its firm texture and rich, almost meaty flavour.

Fresh *shiitake* can be grilled, braised, stir fried or added to stews and one-pot dishes; they can also successfully be made into European-style mushroom soup. The coarse stem should be discarded and the caps carefully wiped with a piece of paper towel or cloth. Do not wash the mushrooms; store them loosely wrapped in paper towel, not plastic, which makes them sweat and decay quickly.

The medicinal benefits of this mushroom, known to the Chinese for centuries, have been confirmed by Western scientists. Fresh or dried, the *shiitake* mushroom lowers cholesterol levels; it also posses anti-viral and possibly anti-cancer properties.

Chinese Spinach

Amaranthus gangeticus;
Amaranthus tricolor

Amaranthus spinach is often regarded as the best of all the tropical spinaches in terms of food value and flavour. Despite the reputed health-giving properties of true or English spinach, Chinese spinach has double the amount of iron and also contains considerable amounts of vitamins A, B and C.

At least seven different cultivars are found in tropical Asia. The most common has pale, almost rounded green leaves, while another variety has dark red markings at the centre of its rounded green leaves. A third variety has darker narrow leaves with pointed tips.

There is no appreciable difference in flavour between these varieties. All should be washed well and the leaves pulled off the stems before cooking. Chinese spinach can be steamed, simmered in soups, or cooked in coconut milk with root vegetables such as sweet potato or pumpkin. If steaming this type of spinach Western style, a few leaves of mint added to the pan improve the flavour.

Botanical Family:
Amaranthaceae

Thai name:
Phak khom suan

Malay name:
Bayam puteh, bayam merah

Indonesian name:
Bayam

Filipino name:
Kulitis

Ceylon Spinach

Basella rubra

Botanical Family:
Basellaceae

Thai name:
Phak prlang

Malay name:
Remayong, Gendola

Indonesian name:
Gendola

Filipino name:
Alugbati

The leaves of this plant, which is widely cultivated in tropical Asia, are used in much the same way as a spinach, hence its name. Despite its name, this vegetable is believed to have originated in India. It has been grown in China for centuries, cultivated mainly for its fleshy red berries which produce a dye used by women as a rouge and by mandarins for colouring seal impressions. It can also be used for colouring jellies or cakes, the colour intensified by the addition of lemon juice.

The distinctive beetroot-red stems and fleshy leaves of Ceylon spinach make it easy to recognise in the markets. It grows easily and keeps well after picking even without refrigeration, provided the stems are put in a jar of water. It has a slightly slippery texture after cooking, and is rich in minerals and vitamins, with mildly laxative properties.

The leaves and tender top portion of the stems are best steamed in the water left clinging after washing; they can also be added to soups and stews.

Papaya

Carica papaya

Papayas are one of the most easily grown and popular fruits found in kitchen gardens throughout the tropics. The unripe green fruit is eaten as a vegetable, particularly in Thailand, the Philippines and Indonesia.

While not as rich in vitamin C as the ripe fruit, green papayas contain an enzyme, papain, which softens meat and is used commercially as a meat tenderiser. Local cooks often mix the pounded leaves or bruised skins with meat to tenderise it. The young leaves of the papaya tree are also edible, but must be boiled in two changes of water.

Shredded green papaya is used for salads, particularly in Thailand, where it is transformed into the famous Som Tom, and in pickles, such as the Filipino Atsara. When cooked, the flesh tastes much like bottle gourd or any other vegetable marrow. It readily takes on other flavours and is even cooked with juniper berries and other seasonings to make a tropical sauerkraut in French Polynesia and the French Caribbean.

Botanical Family:
Caricaceae

Thai name:
Malakor

Malay name:
Buah betik

Indonesian name:
Pepaya

Filipino name:
Papaya

English Spinach

Spinacea oleracea

Botanical Family:
Chenopodiaceae

Thai name:
Phak puay-leng

Malay name:
Bayam po choy

Indonesian name:
Peleng, puileng

Filipino name:
Kulitis

The origin of this plant is unclear, but it is believed to have spread from Southwest Asia to both Europe and China. Although there are a number of tropical leafy greens loosely referred to as "spinach", this is the only true spinach and has an incomparable flavour. It is quite expensive, and much prized by the Chinese.

As this vegetable prefers a cooler climate, it is grown only in hilly areas in tropical Asia. The leaves bruise easily, so care must be taken when choosing spinach to ensure the leaves are not already damaged or they will perish quickly. Store wrapped in a paper towel for a day or so and wash thoroughly before using both the tender stems and the leaves.

Often called "Cameron Highlands Spinach" (referring to where it is grown), this spinach is also known by its Cantonese name, *poh choy*. It can be cooked in any style, Asian or Western, but benefits from only subtle seasoning to allow its own flavour to dominate.

Lettuce

Lactuca sativa

Lettuce is a vegetable which has been cultivated for thousands of years, and is believed to be derived from a wild lettuce indigenous to Western Asia. In China, lettuce has been grown for more than a thousand years. One particular variety (stem or asparagus lettuce) is grown for its thick edible stem. It is sometimes pickled and canned, a somewhat surprising concept to those accustomed to lettuce as a salad vegetable.

Two varieties of lettuce are found in tropical Asian markets. When the vegetable is grown in the lowlands, it has long loose leaves with frilled edges, and does not form a tight round head like the common iceberg variety. Loose-leaf lettuce (often termed "local lettuce") lacks the crispness and flavour of the round variety. It is often used as a garnish or else added, both leaves and sliced stems, to soups in most of Southeast Asia.

The round or "head" lettuce found in temperate climates is grown in highland areas of tropical Asia.

Botanical Family:
Compositae

Thai name:
Phak khat horm

Malay name:
Selada

Indonesian name:
Salada

Filipino name:
Letsugas

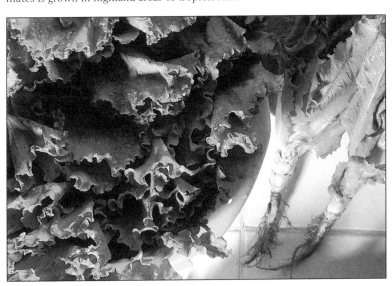

Chrysanthemum

Chrysanthemum coronarium

Botanical Family:
Compositae

Thai name:
Phaktang-o

Malay name:
Tungho

Indonesian name:
Tungho

Filipino name:
Tunghao

Everyone is familiar with the brightly coloured chrysanthemum flowers, native to Europe and north Asia, but not many are aware that the leaves of a certain variety of chrysanthemum are edible. Do not be tempted, however, to cook the leaves of the next bunch of chrysanthemums you buy from the florist; they may technically be edible, but they certainly won't be nice.

In tropical Asia, the tender leaves of a particular variety known as the garland chrysanthemum are eaten mainly by the Chinese, who add small amounts to soup or sometimes cook them as any other leafy green vegetable. They seem to be particularly popular in the hot pot or fondue known in Malaysia and Singapore as "steamboat".

The leaves can also be dipped in batter and deep fried (a popular Japanese treatment). Perhaps because of its pungent smell, the garland chrysanthemum is not widely eaten, although it is rich in vitamins A and B.

Taro

Colocasia esculenta

The taro, also known as cocoyam or dasheen, is sometimes confused with the yam. To add to the confusion, sweet potatoes are often referred to as yams in the USA. Taro is superior in flavour and texture to the yam, which is more popular in the Pacific than in tropical Asia. The starchy paste known as *poi* eaten in Hawaii and other parts of the Pacific is made from taro.

There are some 200 varieties of taro, which is sometimes referred to as "the potato of the tropics", although sweet potatoes could also make this claim. The pinkish-white fleshed taro, recognised by the ring of colour at the base of the stem, probably has the best flavour and texture.

The bulging starchy corms can be roasted, fried, boiled and mashed to form croquettes or grated to form a taro "basket" (a Cantonese restaurant favourite). The young leaves can be cooked in the same way as any other leafy green. Taro corms contain calcium oxalate crystals, so must always be boiled to destroy these.

Botanical Family:
Araceae

Thai name:
Phueak

Malay name:
Keladi, talas

Indonesian name:
Keladi

Filipino name:
Gabi

Sweet Potato

Ipomoea batatas

Botanical Family:
Convolvulaceae

Thai name:
Man thet

Malay name:
Ubi keledek

Indonesian name:
Ubi manis, ubi jalar

Filipino name:
Kamote

The root tubers of this tropical American native are very popular in many parts of Southeast Asia, and as the plant grows easily, it is often planted in home gardens. Both the young leaves and the tubers can be eaten, the former simmered in soups, stir fried or stewed, the tubers cooked in a variety of ways.

There are several different shapes, sizes and colours of sweet potato. A purple-fleshed variety is particularly popular in the Philippines. Malaysians and Indonesians generally prefer the bright yellow or orange-fleshed sweet potato to that with white flesh; interestingly, the former has a higher content of vitamin A.

Although peeled chunks of sweet potato are often cooked in coconut milk with leafy greens, the flesh is also boiled and mashed to make a number of savoury snacks. Diced sweet potato is also used in desserts, particularly with sweetened coconut milk; cubes of yam, slices of banana and sago balls are often added to the concoction.

Water Convolvulus

Ipomoea aquatica

There seem to be more English names for this leafy green than almost any other tropical vegetable: water convolvulus, water morning glory, water spinach and swamp cabbage being the most common. It usually thrives in marshy ground, although one variety is grown in normal seed beds like other vegetables. It is a member of the same family as the common morning glory, whose purple, pink or white flowers can be seen growing wild in much of Asia.

Water convolvulus is very rich in iron and vitamin A, and lacks the bitterness of some other iron-rich greens. It is, in fact, one of the nicest leafy greens grown in the tropics. The tender shoots are eaten raw in Thailand, although in most other areas of Southeast Asia, the shoots and leaves are stir fried or braised. The Filipinos make a type of pickle from the hollow stems.

At least two varieties of water convolvulus are cultivated, the one with a slender, blade-like leaf generally being regarded as superior in flavour.

Botanical Family:
Convolvulaceae

Thai name:
Phak bung

Malay name:
Kangkung

Indonesian name:
Kangkung

Filipino name:
Kangkung

Flowering Cabbage

Brassica chinensis var. *parachinensis*

Botanical Family:
Cruciferae

Thai name:
Phak kwaang tung

Malay name:
Sawi

Indonesian name:
Sawi

Filipino name:
Petsay

This is a variety of the popular white cabbage or bok choy. As the English name hints, this particular Chinese cabbage has yellow-flowered shoots when mature, although it is usually sold before these are evident. The leaves are less tightly packed than with many other varieties of cabbage. Both the leaves and stems are eaten, while the flowering tips are also edible.

Many Asians regard the flowering cabbage as the best of all Asian varieties, and it is very popular with Chinese cooks (the Cantonese name is *choy sam*). Flowering cabbage is often cut in 5–8 cm (2–3 inch) lengths and added to soup and noodle dishes. In such cases, the cabbage is often briefly blanched before being added, although it can also be stir fried or braised without this step.

Take care not to overcook flowering cabbage, as its crisp texture and bright green colour are part of its appeal. Stir frying is recommended as this method tends to conserve the vitamins and minerals.

Celery Cabbage

Brassica pekinensis var. *cylindrica*

Also known as long white cabbage, this vegetable is indeed long, although very pale greenish yellow rather than white in colour. Celery cabbage is almost cylindrical in shape, with the leaves tightly overlapping, as with the common round cabbage found in the West. There is some variation in size and shape, with a long slender cultivar as well as one which is almost squat and fat.

This cabbage is native to north Asia, as might be guessed from the translation of its botanical name, Peking Cabbage. It is one of the most popular vegetables in northern China, and is also the basis of the fiery, garlic-laden *kim chee* pickle which is an indispensible part of the Korean diet. The vegetable is now grown in many countries in Southeast Asia, Europe and the USA.

Celery cabbage keeps well even without refrigeration if it is hung up in an airy place. It is excellent eaten raw, and can also be stir fried. If braised or slow cooked in stews or soups, the leaves exude quite a lot of moisture.

Botanical Family:
Cruciferae

Thai name:
Phak kaet khaao-plee

Malay name:
Kubis cina

Indonesian name:
Sawi putih

Filipino name:
Petsay tsina

White Cabbage

Brassica rapa subsp. *chinensis*

Botanical Family:
Cruciferae

Thai name:
Phak kaet bai

Malay name:
Sawi putih

Indonesian name:
Sawi, petsai

Filipino name:
Pechay

There are so many different cultivars of *brassica* or cabbage grown commercially in Asia that identification is often difficult. Even the botanists find this a challenge, and resort to classifying them on the basis of chromosome numbers; the fact that these cabbages can easily be crossbred only adds to the complexity.

White cabbage is one of the most popular of the 40 or so commonly cultivated species, and is often referred to abroad as bok choy (from the Cantonese *pak choi*). Despite its common English name, only the stems of this cabbage are white, the elongated leaves being either bright or dark green. A small variety of white cabbage, often no larger than 12 cm (5 inches), has light green stems; it is usually braised whole, or halved lengthwise.

When dealing with the larger white cabbages, pull the leaves off the central stem and chop them coarsely before cooking. The leaves are also salt-pickled and used in soups or stews in Chinese cuisine.

Mustard Cabbage

Brassica juncea var. rugosa

There are at least eight cultivars of this cabbage, two of which are are commonly available in the markets. The first, wrapped heart mustard, consists of relatively large heads of large flat leaves with wrinkled edges. This variety is particularly pungent, so is never eaten raw. The individual leaves are pulled off, cut and added to soups or stir fried.

Heart mustard is even more popular salt-pickled and is sold from big ceramic jars in many markets. Salted heart mustard cabbage must be rinsed before being added to soups, or finely chopped and used as a flavouring in other dishes.

The other common cultivar of mustard cabbage, bamboo mustard, is much more slender and does not form a tightly wrapped head. As it lacks the bitterness of the heart mustard, it is commonly eaten fresh. One popular Cantonese way of using bamboo mustard is to trim off and discard the leaves. The wide stem is then stuffed with a prawn or crab paste, and steamed.

Botanical Family:
Cruciferae

Thai name: Phakkaat-khieo

Malay name:
Sawi

Indonesian name:
Sawi hijau, moster

Filipino name:
Mustasa

Watercress

Rorippa nasturtium-aquaticum

Botanical Family:
Cruciferae

Thai name:
Phak kaat nam

Malay name:
Selada air

Indonesian name:
Selada air, kenci

Filipino name:
Watercress

The crisp, slightly peppery leaves and young shoots of watercress have been enjoyed raw as a salad "since time immemorial", according to the famous botanist, G. A. C. Herklots. The plant grows wild in the waterways of Europe, the USA and New Zealand, and has been cultivated in tropical Asia for the past century, where it is grown during the cool season or at higher altitudes year round. In some highland areas of tropical Asia, it can even be seen growing wild at the edge of canals or drains.

Although watercress is commonly eaten raw in Western countries, it is always cooked in Asia, generally in the form of soups. The Chinese believe that watercress is "cooling" to the system, and that is is also mildly laxative.

. When choosing watercress in the market, make sure it is bright green with no yellowish leaves. Watercress is highly perishable and can be stored for only 24 hours or so in the refrigerator; discard any yellowish leaves and use the leaf sprigs and tender stem tips.

Giant White Radish

Raphanus sativus

Because of their similarity in appearance—if not in colour—the Chinese refer to the giant white radish as "white carrot", while the carrot is known as "red carrot".

White radishes found in Southeast Asia can be anything from 12–36 cm (5–15 inches) long, considerably smaller than their cousins from cooler north Asia. White radish is particularly popular in Japan.

Giant white radish can be cut in chunks and cooked in stews, where it loses most of its pungency, and makes a passable substitute for turnip in Western dishes. In the tropics, however, it is more popular as a refreshing salad or fresh pickle.

To reduce some of the "bite", radish eaten raw is generally sliced or grated, sprinkled with salt and left to stand for about 10 minutes. It is then washed, drained and combined with a dressing. The Chinese often combine strips of white radish with half-ripe papaya to make a popular appetiser served in restaurants.

Botanical Family:
Cruciferae

Thai name:
Hua phak kaat

Malay name:
Lobak

Indonesian name:
Lobak

Filipino name:
Labanos

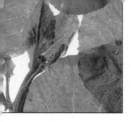

Chinese Kale

Brassica alboglabra

Botanical Family:
Cruciferae

Thai name:
Khana

Malay name:
Kai lan

Indonesian name:
Kai lan

Filipino name:
Broccoli

Sometimes referred to as Chinese broccoli, this vegetable is more commonly known in Asian markets by its Cantonese name, *kai lan*. Believed to be a variety of the European kale or *B. oleracea*, Chinese kale is particularly popular with Chinese cooks throughout Southeast Asia, and also with the Thais.

The leaves are generally pulled off the central stem, and the stems (except for the tough end) peeled and sliced for stir frying. The stems have a better flavour and texture than the tough, somewhat bitter leaves; some cooks discard all the leaves, while others use only the inner, tender leaves.

Relatively recently, a sweet-tasting "baby" Chinese kale has appeared in many markets; only 5–8 cm (2–3 inches) in length, it is grown by crowding the *kai lan* seedlings and growing them with liberal amounts of fertiliser. Baby Chinese kale braised with oyster sauce is currently one of the most popular vegetable dishes in Chinese restaurants.

Chayote

Sechium edule

This vegetable, native to Southern Mexico and Central America, grows prolifically when planted in hilly regions of tropical Asia. It is fairly neutral in taste and rather poor in vitamin and mineral content. Perhaps it is the versatility of the pear-shaped fruit, known as choko in Australia, which makes it popular in some countries, particularly the Philippines.

Avoid any chayote which is starting to sprout. Peel the somewhat prickly, wrinkled skin and either halve or slice the chayote.

Simmer or add to other vegetables in a stew; don't discard the seed—when boiled, it is not only edible but delicious. Halved chayote can be filled with minced meat and steamed; it can also be steamed and then used to hold a savoury filling, in the way avocadoes are often filled with prawns. Raw chayote can be diced and used in salads like celery, or can be added to apples (though it takes longer to cook) to pad out an apple pie.

Botanical Family:
Curcurbitaceae

Thai name:
Ma-kheua-krena

Malay name:
Labu Siam

Indonesian name:
Labu Siam

Filipino name:
Sayote

Angled Gourd

Luffa acutangula

Botanical Family:
Cucurbitaceae

Thai name:
Buap liam

Malay name:
Ketola, petola sagi

Indonesian name:
Belustru

Filipino name:
Patola

It's easy to see why this gourd got its common English name, as it has ten angular ridges running from the stalk to the tip of the fruit. Choose young fruit, which are smaller in size, as the interior becomes tough and fibrous in mature-angled gourds. Peel off the ridges and the rest of the skin, which is bitter. Slice crosswise, leaving the seeds intact, and use in soups or stir-fried dishes.

A very similar gourd, sometimes known as sponge gourd (*L. cylindrica*), lacks the ridges of the angled gourd. It is peeled and eaten in much the same way as the angled gourd, but is also grown for another important reason: the fibrous interior of the dried fruit is the source of the bathroom loofah.

Sponge gourds are also commonly used by Asian housewives in the villages to clean utensils and pots (hence its other name, dish rag gourd). Sponge gourds also have a number of industrial uses, including being used as filters to remove oil from water.

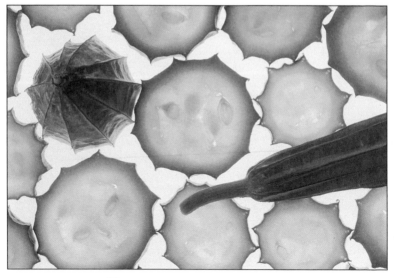

Bitter Gourd

Momordica charantia

Also known as bitter cucumber, bitter melon or balsam pear, this vegetable is, predictably, very bitter in flavour. It grows easily and the climbing vine is often seen in kitchen gardens. The fruit looks rather like a pale green cucumber with a bumpy, grooved skin.

    To avoid excessive bitterness, the bitter gourd is picked while young and is usually rubbed with salt and left to stand for about an hour before cooking; it is then rinsed and dried, and the central spongy portion and seeds discarded.

    Thai and Chinese cooks like to stuff thick slices of this gourd with minced pork; it is also popular fried with eggs. Other cooks in tropical Asia sometimes skip the preliminary salting and instead simmer slices or cubes in salty water before adding them to soups or a seasoned stew.

    The bitter gourd is rich in vitamins A and C. Perhaps because anything bitter is thought to be good for you, many Asians believe this vegetable has medicinal properties.

Botanical Family:
Cucurbitaceae

Thai name:
Mara

Malay name:
Peria

Indonesian name:
Peria, pare

Filipino name:
Ampalaya

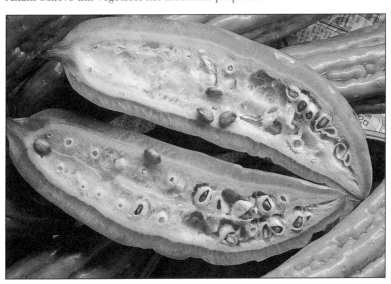

Bottle Gourd

Lagenaria siceraria

Botanical Family:
Cucurbitaceae

Thai name:
Namtao

Malay name:
Labu air

Indonesian name:
Labu air

Filipino name:
Upo

This gourd is believed to be native to Africa, and since experiments show that mature gourds have floated on the sea for more than 224 days without detriment to the seeds, it is quite possible they were distributed naturally.

The bottle gourd is an ancient vegetable, the dried skins having been unearthed in Egyptian tombs dating back to 3500 BC. There is evidence of its existence in Peru and Mexico even earlier than that.

The vegetable is picked for eating while still slender and pale green. This is one of the best gourds, with firm yet tender flesh and a mild flavour. It is popular throughout Asia, particularly with Southern Indian cooks, who transform it with spices, grated coconut or yoghurt, while Malay cooks prefer to add it to coconut milk curries.

If the gourd is left to mature and dry, it forms a tough woody shell that can be used as a storage container or bottle, cut to form bowls or cups, shaped into a ladle or made into a musical instrument.

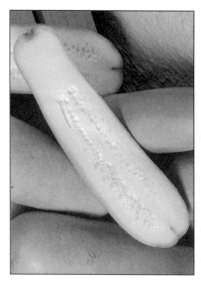

Snake Gourd

Trichosanthes cucumerina var. *anguina*

The snake gourd is sometimes confused with a snake-like variety of the bottle gourd, although if you see a long twisting gourd with streaks of darker green showing through the chalky white skin in the market, it's likely to be a snake gourd.

This particular plant is of Indian origin, and is thus very popular with Indian cooks throughout Southeast Asia. The gourd grows naturally in a twisted fashion; quite often, however, a stone is tied to the end of the young fruit to help it grow straight (and, presumably, longer).

Snake gourds must be eaten when young. The skin is usually rubbed with coarse salt to remove the waxy white surface, which develops naturally in all snake gourds. The vegetable is then rinsed, cut into pieces and simmered in curries and stews. Because the snake gourd has a bitter flavour, it is regarded as a tonic and is used in traditional medicine. It is, incidentally, more nutritious than most other gourds.

Botanical Family:
Cucurbitaceae

Thai name:
Buap nguu

Malay name:
Petola ular

Indonesian name:
Paria belut

Filipino name:
Papukis

Winter Gourd

Benincasa hispida

Botanical Family:
Cucurbitaceae

Thai name:
Fak nguu

Malay name:
Kundur

Indonesian name:
Kundur, baligo

Filipino name:
Kondol

Two varieties of the winter or wax gourd are found in the markets, differing so much in appearance that it is surprising to learn they are the same species. The most spectacular is the round variety, often of huge size, which has a waxy surface and seeds embedded in the pulp in the centre of firm, whitish-green flesh.

Sometimes called winter melon, this gourd is a favourite for soup, especially in Chinese restaurants where skilled chefs carve dragons or other fanciful motifs into the outside of the skin, and use the gourd as a soup tureen. It is often possible to buy a portion of the large winter gourd to dice and add to soups.

A much smaller, elongated variety of winter gourd lacks the waxy exterior of its round cousin, and is instead covered with very fine silky white hairs. It can be peeled and eaten raw, although is more commonly cooked in soups or stews. It is also popular hollowed, stuffed with a savoury meat filling and steamed or added to soups.

Pumpkin

Cucurbita moschata

Pumpkin or winter squash, as the Americans call it, is a native of North and South America and is popular throughout most of tropical Asia today. There are many varieties, the most common having beige-brown skin with golden or orange flesh and a pulpy interior filled with seeds. These seeds are dried and eaten as a nutritious snack, especially among the Chinese.

Whole pumpkin keeps reasonably well even in the humid tropics, although once it is cut, it will start to deteriorate after a few days.

The flesh is cooked in a variety of ways, in soups, curries and in coconut milk. It is not only treated as a vegetable but also used in desserts, especially in Thailand (where tiny pumpkins are steamed with a rich coconut custard filling), and Malaysia and Indonesia, where it is simmered with coconut milk and sugar. The tips of the pumpkin vine are eaten as a vegetable, and the flowers, which can be stuffed, are found in some markets.

Botanical Family:
Curcubitaceae

Thai name:
Fak thong

Malay name:
Labu merah

Indonesian name:
Waluh, labu merah

Filipino name:
Kalabasa

Water Chestnut

Eleocharis dulcis

Botanical Family:
Cyperaceae

Thai name:
Haeo cheen

Filipino name:
Apulid

The grass-like sedge growing in swampy ground in China seems like an unpromising candidate for producing the crisp, sweet corms known as water chestnuts. The round brown-skinned corms have a flattened top and base. When peeled (time-consuming, but worth the effort), the water chestnuts reveal their bright white flesh.

Popular primarily with Chinese and Thai cooks, water chestnuts are found in markets throughout the cities of Southeast Asia, as well as in some provincial markets. Choose carefully, rejecting any corms which feel soft. Always take a few more than needed as almost invariably there will be several which have turned yellowish-brown inside and must be discarded.

Water chestnuts can be eaten both raw and cooked, and are often finely diced and added to stuffings mainly for their crisp texture. The flavour of water chestnuts is delicious and somewhat sweet, hence the botanical description, *dulcis*.

Cassava (Tapioca)

Manihot esculenta

A native of Central or South America, cassava has been successfully introduced throughout Asia and the Pacific and is now an important crop. It grows easily in almost any soil, and even in extremely dry conditions. Often regarded as a subsistence crop, cassava is not always available in big city markets.

Cassava must be cooked to destroy the hydrocyanic acid which it contains. The tender young leaves from the top of the cassava plant are boiled and eaten as a green vegetable in much of tropical Asia. The mature starchy root tubers are boiled to make a potato-like staple, while the young roots are peeled and grated to make various cakes and savouries. Unfortunately, most of the cassava sold in the markets is too old for this.

Those familiar with the small dried white balls similar to pearl sago and sold as tapioca in supermarkets and grocery stores may be surprised to learn that they are made from the starch of the cassava or tapioca plant.

Botanical Family:
Euphorbiaceae

Thai name:
Man sampalang

Malay name:
Ubi kayu

Indonesian name:
Ubi kayu, singkong

Filipino name:
Kamoteng-kahoy

Sayur Manis

Sauropus androngynus

Botanical Family:
Euphorbiaceae

Thai name:
Phak waan baan

Malay name:
Sayur manis, cekur manis

Indonesian name:
Katuk

Filipino name:
Binahian

There appears to be no common English name for this vegetable, except in Singapore and Malaysia where a tender variety of sayur manis is often called "Sabah vegetable", referring to the fact that this particular type was developed in the north Borneo state of Sabah.

A small shrub found growing wild in much of tropical Asia, sayur manis produces leaves rich in vitamins A and C. The tender tips are picked off and braised, cooked with eggs or added to soups. It was once a great standby in local gardens, constantly producing a crop of edible leaves.

Sabah vegetable, the relatively new type of sayur manis with stems which resemble asparagus in texture, has an excellent flavour. It was developed by force-feeding the cultivated plant to encourage it to grow rapidly. The original sayur manis plant has tough woody stems and only the leaves are eaten. The popularity of Sabah vegetable seems to be spreading and it is now being grown in other areas of Malaysia.

Bamboo Shoot

Bambusa sp., *Dendrocalamus* sp.,
Phyllostachys sp.

Tender young shoots from many species of bamboo are eaten. Some are infinitely superior, lacking the bitterness of others; bamboo shoots from a cooler climate including those grown at higher altitudes are generally much sweeter. Only when you have tasted succulent fresh winter bamboo shoots or one of the finest high-altitude varieties can you understand why Chinese poets waxed so lyrical over this vegetable.

Bamboo shoots are sold in several ways in the markets. They may be whole, complete with their leaf sheaths; still whole but peeled; peeled and sliced ready for cooking or cut and pre-boiled.

Before cooking whole bamboo shoots, remove the sheaths and cut the shoot in half lengthwise. Boil in unsalted water for 20–30 minutes. Drain and cut in chunks, crosswise slices or shreds according to your recipe. Bamboo shoots are commonly used for stir frying, soups, braising or curries.

Botanical Family:
Gramineae

Thai name:
Nor mai pai tong

Malay name:
Rebung

Indonesian name:
Rebung

Filipino name:
Labong

Corn

Zea mays var. *rugosa*

Botanical Family:
Gramineae

Thai name:
Khaao phot on

Malay name:
Jagung

Indonesian name:
Jagung

Filipino name:
Mais

Maize, one of the many tropical American natives now found virtually worldwide, is grown primarily as stock feed in Asia. However, a sweet-tasting hybrid known as sweetcorn (*Zea mays* var. *rugosa)* is often found in the markets and, when roasted or boiled, is a favourite street-stall snack.

Miniature corn, harvested when only 8–10 cm (3–4 inches) long, has been popular in Thailand for a number of years and is now being cultivated in several other areas of Southeast Asia. Experts differ as to whether this miniature corn is from a special dwarf variety or whether it is a new variety which has crowded side "ears". Be that as it may, the entire baby corn cob is delicious either raw or cooked. It can be added to soups or stir-fried vegetable combinations, or simply enjoyed on its own.

Interestingly, sweetcorn is often used in desserts and cakes in much of Southeast Asia, and is even found as a flavouring for ice-cream.

Long Bean

Vigna sesquipedalis

Also known as cow pea or, in the USA, as yard-long bean, this plant is probably native to China. The long bean has been cultivated since ancient times, and several varieties are found in Asian markets. The most common has smooth pale green pods, which are up to 40 cm (16 inches), well short of a yard. A similar variety is slightly more slender and a darker green, with a more intense flavour. Sometimes, another variety—deep brownish red in colour—can also be found.

Young long beans are commonly served in a platter of mixed vegetables which usually includes round cabbage and sprigs of water convolvulus, and eaten raw with a spicy dip in Thailand. In Malaysia and Indonesia, cooks prefer to blanch the beans first then add them to mixed vegetable salads. Long beans can also be stir fried or cooked with coconut milk. Unlike the common (French) green bean, long beans do not have "strings" and need only to have the stem end cut off before cooking.

Botanical Family:
Leguminosae

Thai name:
Thua fak yao

Malay name:
Kacang panjang

Indonesian name:
Kacang panjang

Filipino name:
Sitaw

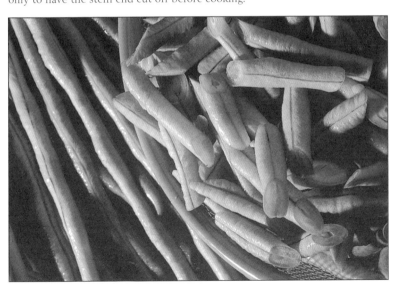

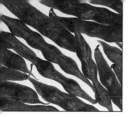

Red-streaked Bean

Phaseolus vulgaris cv.

Botanical Family:
Leguminosae

Malay name:
Kacang merah

The bean family has several hundred varieties, which makes classification somewhat confusing for the average person. Furthermore, the term "bean" is used both for fresh beans eaten in the pod and for the dried seeds of many varieties of bean.

Beans, which are native to Central and South America, have been used as food for literally thousands of years and were brought to Europe by Spanish explorers in the 16th century.

In recent years, a new variety of bean has been cultivated in parts of Southeast Asia, particularly in Malaysia where it is grown in highland areas. This bean, known in some European countries by its Italian name, borlotti bean, has creamy-coloured pods streaked with a pinkish-purple colour. The pods are picked when the seeds are mature but not yet dried, and the seeds of "beans", which have the same streaked appearance as the pods themselves, are simmered, generally in soups.

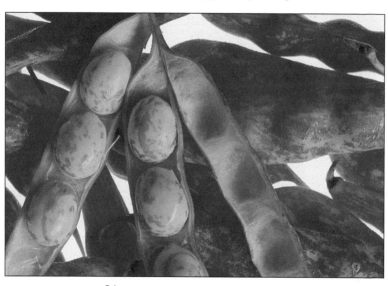

Winged Bean

Psophocarpus tetragonolobus

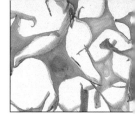

When you look at this bean, it's easy to understand why it is known as winged bean or four-angled bean: the pale green pods, which are about 12–15 cm (5–6 inches) long, have a frilly edge along four sides. In certain areas they are known as Goa bean, and, as the name indicates, some authorities suggest their origin was India.

Cultivated all the way from India to New Guinea today, the winged bean is highly nutritious, containing both protein and oil (especially in the ripe seeds). Unless picked very young (which you can only be sure of if you grow them yourself—which is very easy to do) winged bean pods have a "string" which should be pulled off. They can then be treated like French beans. They are popular in salads in Thailand, where they are eaten raw or blanched.

The pretty white or pale blue flower, young shoots and tuberous roots are all edible; if they are available, why not add a few fresh flowers to a salad?

Botanical Family:
Leguminosae

Thai name:
Thua phuu

Malay name:
Kacang botor, kacang kelisa

Indonesian name:
Kecipir, kacang botor

Filipino name:
Sigarilyas

Twisted Cluster Bean

Parkia speciosa

Botanical Family:
Leguminosae

Thai name:
Sator

Malay name:
Petai

Indonesian name:
Petai

Very different from the average vegetable, these beans grow on a huge tree found wild in the forests of Malaysia, Indonesia and Thailand. The beans grow in clusters of several pods, with the swollen seeds clearly visible through the bright green pod. The pod, which contains about 10–18 large seeds or beans, is opened by pulling the "strings" off each side and twisting.

The beans themselves should be peeled, a somewhat troublesome task unless they are at the peak of ripeness. Some markets sell the beans already peeled and packed in water inside plastic bags, but check they are still fresh by sniffing them to be sure they have not fermented.

Twisted cluster beans have a very strong smell and bitter taste; like the durian fruit, some find it offensive, others love it. Like asparagus, the odour of this vegetable is present in urine passed after eating. The peeled beans are generally blanched and eaten with a dip, or cooked in spicy sauce; they seem to go well with prawns.

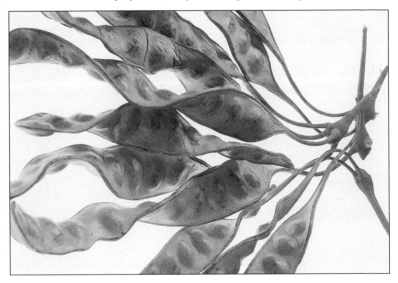

Beansprouts

Vigna radiata & Glycine max

The most common variety of beansprouts are produced by soaking dried mung beans and keeping them moist for three to four days until a crisp white shoot sprouts. Beansprouts are most commonly associated with Chinese cuisine, but the mung bean is actually a native of India.

Soybean sprouts, produced from the much larger dried soybean, are larger and fatter than the more common mung bean sprouts, and are even more nutritious.

Beansprouts can be refrigerated for several days if the water is changed daily. If you buy sprouts which have already had the straggly roots pinched off (an aesthetic touch which is not really necessary), use them on the day of purchase as they spoil more quickly. Sprouts are eaten raw, slightly blanched or stir fried for a few seconds so they retain their crunchy texture. Sprouts should be well drained, and if stir fried, should be cooked over very high heat. Soybean sprouts take slightly longer to cook than regular beansprouts.

Botanical Family:
Leguminosae

Thai name:
Thua ngok

Malay name:
Taugeh, kacang cambah

Indonesian name:
Taoge, kecambah

Filipino name:
Toge

Snow/Sugar Pea, Pea Shoot

Pisum sativum var. *saccaratum*

Botanical Family:
Leguminosae

Thai name:
Thua lantao

Malay name:
Kacang polong

Indonesian name:
Kacang ercis, kacang kapri

Filipino name:
Sitsaro

The common Western green pea pods which provide tender peas inside inedible pods are not found in tropical Asian markets. Instead, there are two other varieties of this ancient vegetable, which probably originated in Southwest Asia.

The more common tropical variety, known as the snow pea, has tiny immature peas inside flat pods. There are two varieties, one about twice the length and width of the other. The "strings" are pulled from the tip of the pod down to the stem end, which is snipped off, and the entire pod used in stir-fried dishes or soups. Snow peas can also be eaten par-boiled and served in salads.

The other variety, sugar pea (also known as honey pea), is very similar to the Western green pea and has larger peas inside the pods than the snow pea. However, unlike Western peas, it is eaten whole, tender pod and all. At Chinese New Year, it is the last ingredient added to the traditional mixed vegetable dish known as *Lo Hon Jai*.

The tender young shoots of both varieties, known as *dou miao*, are an expensive and highly esteemed leafy green favoured by Chinese cooks. They comprise a short stem bearing pairs of small, oval-shaped leaflets. They grow as creepers and are prevented from flowering or fruiting. Because they lack mature fibre, they wilt easily and often look rather sad sitting in heaps in the market.

Pea shoots can be prepared in a number of different ways. They are usually (due to the price) served only on special occasions. They can be par-boiled in soup, or simply stir fried with fresh ginger. A well-known restaurant dish is *dou miao* stir fried with crab meat. It is sometimes possible to buy the young shoots when they are really tiny, and in this condition, they may be added raw to leafy green salads or added at the last moment when serving soups.

Because they deteriorate quickly, store only for 24 hours, loosely tossed in a container in the refrigerator.

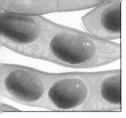

Soybean

Glycine max

Botanical Family:
Leguminosae

Thai name:
Thua rae

Malay name: Kacang,
soya putih

Indonesian name:
Kacang kedelai

Filipino name:
Utaw

The soybean has been cultivated in China for around 3,000 years, and in recent times has been hailed as an extremely important crop for both human consumption and animal feed. Containing as much as 40% protein, dried soybeans are particularly popular with vegetarians. Many important Asian foods are made from soybeans, including soy sauce, soybean curd and soybean milk.

Although the immature green pods containing tender fresh beans are highly prized in Japan, Taiwan and parts of China, they are not commonly seen in tropical Asia, even though they are grown in some countries in the area.

If you are fortunate enough to find fresh soybean pods, which have a fuzzy skin and contain three or four beans inside, steam or boil them in salty water. The only practical way to eat them is to pick up the pods with your fingers, suck them to extract the soft portion inside the tough skin, and then squeeze out the beans. Frozen soybean pods are generally sold in Japanese supermarkets.

Yam Bean

Pachyrrhizus erosus

The name of this Central American native is somewhat confusing, as it is most definitely not related to true yams and the portion that is eaten is not a bean.

A beige-coloured skin—which is obligingly easy to peel off—surrounds the crisp white flesh of this edible root tuber. The flavour of the crisp-textured yam bean has been described as a cross between an apple and a potato. It is particularly popular in the Philippines (where it was introduced by the Spanish), and when young and sweet, is sold as a street-side snack.

Although excellent raw (especially when dipped into a spicy *sambal* or *phrik)*, the yam bean can be cooked. It is often grated and simmered together with more expensive bamboo shoots as a filling in Malaysia and Singapore (where it is sometimes misleadingly called a turnip).

An increasingly popular international name for the yam bean, *jicama*, comes via the USA from Mexico, where it is generally boiled.

Botanical Family:
Leguminosae

Thai name:
Man kaeo

Malay name:
Sengkuang, bengkuang

Indonesian name:
Bengkuang

Filipino name:
Singkamas

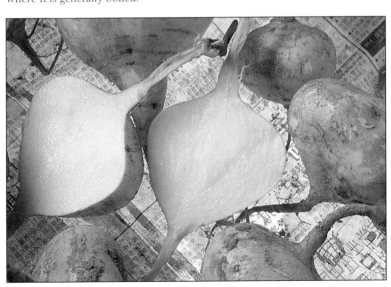

41

Chinese Chives

Allium tuberosum

Botanical Family:
Liliaceae

Thai name:
Kui chaai

Malay name:
Kucai

Indonesian name:
Kucai

Filipino name:
Kutsay

Also referred to as garlic chives or flat chives, this plant is composed of clusters of long, flat leaves about 0.25 cm ($^1/_4$ inch) wide. As the name "garlic" suggests, this plant is actually a variety of *Allium*, but the only similarity it has with true garlic is the pungent flavour of its leaves.

Chinese chives are grown for two reasons. Cooks use the leaves for stir frying as a vegetable or in the same way as Westerners use chives. Or they use the stems with the edible flowering culm at the tip for their onion flavour, again in stir-fried dishes or as a side dish with seafood. When found in the flowering latter stage, they are considerably more expensive.

A creamy white variety of garlic chives, produced by growing the plant under cover, is sometimes found in Asian markets. There is no appreciable difference in flavour, but the tender texture and unusual colour seems to appeal to Chinese cooks, the main market for this high-priced "albino".

Garlic

Allium sativum

It is difficult to imagine any Asian market without garlic, an essential ingredient in every regional cuisine. The medicinal value of garlic (which is known to have, among other attributes, antibiotic properties) was acknowledged back in ancient times, with records of its use in both Sanskrit and Egyptian literature. It is believed that the frequent consumption of garlic in the tropics may reduce the incidence of intestinal disorders.

A head of garlic is made up of clusters of "cloves" or "pips", the size of which varies from one Asian country to another (the smallest must be those found in the Philippines). The amount used depends on personal taste, with raw garlic being much more pungent than "cloves" which have been slowly cooked for a considerable period of time. Although almost always used as a seasoning, garlic is treated as a vegetable in one southern Indian curry.

Garlic sold in the markets is already dried, and can be stored in a dry place for several weeks.

Botanical Family: Liliaceae

Thai name: Krathiam

Malay name: Bawang putih

Indonesian name: Bawang putih

Filipino name: Bawang

Spring Onion

Allium fistulosum

Botanical Family:
Liliaceae

Thai name:
Ton horm

Malay name:
Daun bawang

Indonesian name:
Daun bawang

Filipino name:
Sibuyas-na-mura

The spring onion, widely used throughout tropical Asia as well as in north Asian countries such as China and Japan, is very popular as a flavouring green or herb, as a garnish and as a vegetable in its own right.

Spring onions, which grow in bunches, have round hollow green leaves tapering to a white stem often covered at the base with an outer layer of purplish skin. They are known by a wide variety of names in Western countries: Welsh onions, Japanese bunching onions, scallions and even shallots (although the true shallot is nothing like a spring onion). The green onion, looking like a much fatter version of the spring onion, grows from another variety (*A. cepa*) and has a stronger flavour.

Spring onions are eaten raw, added to soups, stir-fried dishes and fillings. One popular Chinese dish stir fries a large bunch of cut spring onions together with cubes of fried beancurd. Spring onions keep well refrigerated if the root ends are put in a jar of water.

Shallot

Allium ascalonicum

Looking like clusters of miniature onions, shallots are covered with a purplish skin which encloses a sweet pinkish interior. The tropical Asian shallot is different in appearance to the European shallot, although the flavour is similar.

Botanists differ as to whether the shallot is a modification of *A. cepa*, the larger brown or purplish-skinned onion, or whether it is a separate species, *A. ascalonicum*. Regardless of classification, the shallot is a popular seasoning vegetable and is also used in pickles. Slices of deep fried shallot are perhaps the most common garnish used by Malaysian and Indonesian cooks.

Shallots contain less moisture than large onions, and are therefore preferred for pounding to make the spice paste (*rempah* or *bumbu*) which forms the basis of many Malaysian and Indonesian dishes. Shallots should be stored in a dry place and can keep for a couple of months in good conditions.

Botanical Family:
Liliaceae

Thai name:
Horm dang

Malay name: Bawang merah

Indonesian name:
Bawang merah

Filipino name:
Sibuyas tagalog

Okra

Hibiscus esculentus;
Abelmoschus esculentus

Botanical Family:
Malvaceae

Thai name:
Kra ciap

Malay name:
Kacang bendi

Indonesian name:
Kopi arab, okra

Filipino name:
Okra

Okra, a cultigen of uncertain origin, is very popular in the eastern Mediterranean, India, Southeast Asia and in the Carribean and the USA, where it is known as gumbo.

The shape and size of okra gives rise to one of its common names, lady fingers (also lady's fingers). In Malaysia and Singapore, where this vegetable is particularly popular with those of Indian origin, it is often called by its Indian name, *bendi*. The green skin of the immature pods is covered with very fine downy hairs. The pods contain many edible whitish seeds and a musculagenous substance which can be useful for thickening a number of dishes.

Look for the smallest, youngest pods in the market; test their freshness by pressing the tip gently; it should snap, not bend. To avoid the stickiness of the okra, cut off the top of the stalk end, leaving the pods whole. Frying rather than simmering okra also reduces the stickiness; in this case, slice the pods coarsely before cooking.

Jackfruit

Artocarpus heterophyllus

Jackfruit, which has a very rich "tropical" odour, grows to an enormous size and ripe fruits can weigh up to 40 kg (88 lbs). Although popular when ripe, this versatile fruit can also be enjoyed as a vegetable in its immature state. Picked when about 25 cm (10 inches) long, the young oblong fruit is covered with green skin which has short bumps. The skin should be cut off with an oiled knife (to avoid problems with the sticky latex) and the interior—excellently flavoured seeds and all—cut into chunks. The flesh is usually simmered in salted water for a few minutes before being drained and cooked with coconut milk.

Two related fruits, the breadfruit (*A. altilis*) and the breadnut, are also treated in the same way. Many markets sell packets of the jackfruit, breadfruit or breadnut already peeled and cut in chunks ready for cooking. Unless you're an expert, it is difficult to tell the three apart; however, it is not really essential as they are cooked in a similar fashion and taste much the same.

Botanical Family: Moraceae

Thai name: Khanun on

Malay name: Nangka

Indonesian name: Nangka

Filipino name: Langka

Horseradish Tree

Moringa oleifera

Botanical Family:
Moringaceae

Thai name:
Phak marum

Malay name:
Kacang kelor

Indonesian name:
Daun kelor

Filipino name:
Malunggay

This Indian native gets its common English name from the fact that its pungent roots were used as a substitute for true horseradish (*Amoracia rusticana*) by the colonial British in India.

The horseradish tree illustrates the point that what people eat is often dictated by custom and not by the actual edibility of a plant or a portion of it. The fern-like young leaves, extremely rich in vitamins A and C, are probably the most common vegetable in the Philippines, where they are added to soups. The leaves are also popular in Thailand, where the young green sprigs of many wild trees are used as vegetables. They are generally blanched and eaten with a spicy dip, or added to soups.

The long, immature seed pods known as drumsticks are considered to have medicinal value. In India they are cut into pieces and simmered (often with *dal*) until tender; only the soft interior is edible. They are also prepared by Malaysian cooks.

Banana Bud

Musa spp.

The tapering, purplish inflorescence hanging at the end of a clump of bananas is normally cut just after the fruit has formed, and is widely used as a vegetable. Certain varieties of banana, such as the plantain or "cooking" banana, produce a better flavoured and textured bud, and these are the ones you're likely to find in the market.

The outer leaf sheaths of the bud should be pulled off, together with the blossoms, until the pinkish-white heart is revealed. Use an oiled knife to cut the bud lengthwise into four; fastidious cooks also pull the hard stamen from the centre of each blossom; this is a fiddly task but worth the effort.

The prepared bud is then chopped and blanched. It is very popular as a salad, with a coconut milk dressing, or eaten with a savoury dip. It can also be simmered in coconut milk after the initial blanching. The flavour is reminiscent of artichokes, and banana bud goes well with a vinaigrette dressing.

Botanical Family:
Musaceae

Thai name:
Hua plee

Malay name:
Jantung pisang

Indonesian name:
Jantung pisang

Filipino name:
Puso ng saging

Lotus Root

Nelumbo nucifera

Botanical Family:
Nymphaeaceae

Thai name:
Rak bua

Malay name:
Teratai, seroja

Indonesian name:
Teratai

Filipino name:
Baino

The lotus is an aquatic plant much prized for the beauty of its flowers and also for its association with Buddha. (The flower, which rises unsullied from a muddy pond, is seen to represent purity.) Indigenous to Asia, all the way from Iran in West Asia to Japan, the lotus has been cultivated in China since the 12th century BC.

Both the rhizome and the seeds of the lotus are edible, although you'll need to have your own lotus pond to enjoy the young green seeds. The rhizome, harvested while still young, is crisp with a delicious flavour. It should be peeled and immediately plunged into boiling water with a few drops of vinegar or lemon juice added to stop it turning brown. Blanch for a few seconds then serve with a tangy dressing, or stir fry. Sliced lotus root can also be fried in batter. Another method is to stuff the holes of a segment of lotus root, slice it, then deep fry the slices or add to soups. Ripe lotus seeds are dried, then boiled and served in syrup.

Capsicum

Capsicum annuum

Also known as bell peppers or sweet peppers, capsicums belong to the same family as the fiery hot chilli and are, like them, native to the Americas. They are grown in elevated areas in tropical Asia, and are popular mainly with Chinese cooks. The capsicum also appears in a number of Filipino dishes of Spanish origin.

Generally much smaller in size than the capsicums grown in Europe, the USA or Australia, tropical capsicums are usually picked unripe (green), although the ripe red capsicum can sometimes be found. The colourful orange, yellow and even purplish capsicums grown in places like Holland are rarer to find in markets.

The Chinese generally stir fry the capsicum with meat and several other vegetables, taking care not to overcook it so that it retains its crisp texture and bright colour. The capsicum is not widely used by other cooks in Southeast Asia, although its increasing cultivation in highland areas may see it gain wider acceptance.

Botanical Family:
Solonaceae

Thai name:
Phrik wan

Malay name:
Lada besar

Indonesian name:
Cabe besar, lombok besar

Filipino name:
Pulang sili

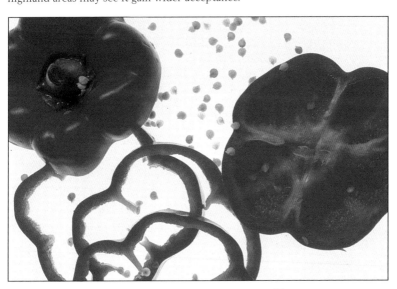

Chilli

Capsicum annuum cv. group *longum*
& *C. frutescens*

Botanical Family:
Solanaceae

Thai name:
Phrik kheefa

Malay name:
Lada, cili, cili padi

Indonesian name:
Cabe, cabe rawit,
lombok

Filipino name:
Sili, siling labuyo

Indispensible throughout tropical Asia today, the chilli is not a native but was introduced from the Americas by the Portuguese and Spanish. Before the advent of chillies, black pepper was used to give a pungent flavour to food. During the past four centuries, the chilli has flourished and today is found in an almost bewildering variety of shapes, sizes and pungencies throughout the region.

Generally speaking, the two most common varieties found in Asia are the finger-length chilli (*C. annuum* cv. group *longum*), sold either green (unripe) or red (ripe), and the fiery little bird's-eye chilli (*C. frutescens*). There are also mild, fat, long chillies; small round chillies; yellow or creamy white chillies; arrow-shaped chillies; pale orange chillies—the range seems endless.

The hottest part of the chilli is the seeds; to reduce the pungency of a dish without losing the chilli flavour, discard some or all of the seeds before using. Take care when preparing chillies as the juice burns sensitive areas.

Eggplant (Aubergine)

Solanum melongena

In a reversal of the more common trend, this is one vegetable native to tropical Asia (India) which was introduced to the Americas. A wide variety of eggplant is found in the tropics; the fruit, regardless of variety, is always much smaller than that of the common European eggplant or aubergine.

The most widespread Asian variety is long and slender, although there are round eggplants ranging in size from the tiny pea-like bitter eggplants (*S. torvum*) which grow wild, up to those the size of a tennis balls. All have a smooth, waxy skin, in colours ranging from deep purple through pale purple, with pale green, white and even bright yellow-skinned varieties.

Tropical eggplants do not, like their Western counterparts, need salting to remove any bitterness. This is a very versatile vegetable, its somewhat bland flavour lending itself to a whole range of spices and seasonings which transform it throughout Asia.

Botanical Family:
Solanaceae

Thai name:
Makhua ling, Ma khua chang

Malay name:
Terung

Indonesian name:
Terung

Filipino name:
Talong

Oyster Mushroom

Pleurotus fiabellatus

Botanical Family:
Tricholomataceae

Thai name:
Hed nang rom
Bhutan

Malay name:
Cendawan

Indonesian name:
Jamur

The Oriental oyster mushroom is sometimes called tree oyster, since it grows on trees (although it can also be grown on logs, straw or sawdust). Several varieties of this mushroom are found, varying slightly in size and colour of the cap. It can be confused with abalone mushroom (*P. cystidiosus*).

The oyster mushroom does not, in fact, taste anything like oysters or abalone. The flavour is mild, lending itself well to partnering with other mushrooms or vegetables. The texture remains pleasantly firm even after cooking.

Oyster mushrooms can be eaten raw in salads, although they have rather a "bite"; in Asia, they are almost always stir fried or braised. They can be used in many different Asian dishes, or cooked with butter and used as a filling for omelettes or as an accompaniment for meat dishes. They are particularly good stir fried with bacon and spring onions, giving a new meaning to the dish "oysters and bacon".

Chinese Celery

Apium graveolens

It is believed that celery is native to both Northern Asia and Europe, and there is evidence that it was known to the ancient Egyptians, Greeks and Romans. The slender plants found in the tropics seem to have originated from celery native to China. It is so radically different in appearance to the large temperate-climate celery that it is hard to believe that it is the same botanical species.

Chinese celery, which is much more pungent than the European plant, has very slender stems and strongly flavoured leaves. It is used primarily as a herb, and only a small amount is needed to add flavour to soups (hence the Malay name, "soup leaf") and stir-fried rice or noodle dishes. The stalks can be finely chopped and mixed with other vegetables, but it must always be remembered that the flavour is very strong.

The large common European celery is found in many city markets in Asia, and is popular for stir-fried vegetable combinations. However, it is not grown in the tropics.

Botanical Family:
Umbelliferae

Thai name:
Phak chee lom

Malay name:
Daun sop, daun seledri

Indonesian name:
Seledri

Filipino name:
Kintsay

Straw Mushroom

Volvariella volvacea

Botanical Family:
Volvariceae

Thai name:
Hed fang

Malay name:
Cendawan

Indonesian name:
Jamur

Filipino name:
Kabute

It is a great pity that this mushroom is so perishable, as it is arguably the most exquisite found anywhere in Asia—except, perhaps, for some of Japan's highly prized wild forest mushrooms. Canned straw mushrooms cannot even begin to approximate the firm, yet yielding texture and delicate flavour of straw mushrooms.

This mushroom, which appears to be native to Asia, is particularly popular in Thailand. Here, the mushrooms are added to soups and chicken curries, or stir fried with prawns and vegetables; Chinese cooks often combine them with crabmeat and egg white sauce.

The straw mushroom can be found wild and is also cultivated on straw or dried vegetable stems. The cap is completely enclosed within its sheath, which eventually splits open to reveal an umbrella-like mushroom inside.

Try to use these mushrooms on the day of purchase, or refrigerate for up to 24 hours loosely wrapped in paper towel.

Ginger

Zingiber officinale

Hundreds of varieties of the *Zingiberaceae* family are found growing wild throughout tropical Asia. The most common, known simply as ginger in English, is cultivated and widely used in cooking. Ginger is also used in the folk medicine of the region, especially for aiding digestion and treating coughs.

The stubby, oblong rhizomes are found in two forms in the markets. Mature ginger, covered with thin beige skin that should be scraped off with a knife before use, is relatively pungent in flavour and preferred for cooking. Two types of mature ginger can often be found in the markets, the fatter variety regarded as better in flavour.

Young ginger, recognised by its delicate skin with the pale lemon colour of the rhizome clearly visible, and by its pinkish-green stems, is less fibrous and milder in flavour. Young ginger is preferred for pickling; it is also grated and squeezed to obtain the juice often used in marinades, especially in Chinese cuisine.

Botanical Family:
Zingiberaceae

Thai name:
Khing

Malay name:
Halia

Indonesian name:
Jahe

Filipino name:
Luya

Vegetable Soup

This is a basic Chinese-inspired recipe which you can adapt according to the availability of ingredients and your taste. Start by making chicken or pork stock, then add the vegetables of your choice, perhaps some cubes of soft bean curd and prawns, and season to taste with a touch of sour tamarind, if liked.

500 g (1 lb) meaty pork bones or chicken carcasses
1 clove garlic, skin left on, lightly bruised
1 sprig of Chinese celery or coriander
1–2 slices fresh ginger
1 litre (4 cups) water
1–2 cups of one or more of the following sliced or chopped vegetables:
    watercress, flowering cabbage, celery cabbage, gourd, spinach,
    giant white radish, lotus root, baby corn, bamboo shoot,
    water convolvulus shoots, sprigs of horseradish leaf
1–2 squares soft bean curd, diced (optional)
100 g (3^1/$_2$ oz) small raw prawns (optional)
salt and light soy sauce to taste
1 tablespoon dried tamarind pulp soaked in ^{1}/$_2$ cup water and
    strained (optional)
sliced spring onion to garnish (optional)
dash of white pepper powder

Put the pork or chicken into a saucepan with the garlic, Chinese celery, ginger and water. Bring to the boil, cover and simmer for 45 minutes. Strain and return stock to the pan, heat and add the vegetables of your choice. Simmer until the vegetables are just cooked, then add the bean curd, prawns (if using), and season with salt, soy and tamarind, if liked.

Reheat and simmer gently for 2–3 minutes until the prawns are just cooked. Sprinkle with spring onion and white pepper and serve hot.

Yam Yai Salad

The Thais probably eat more raw vegetables than anyone else in tropical Asia. A salad may be as simple as a few young long beans, shoots of water convolvulus and chunks of round cabbage dipped in a spicy hot sauce, or it can be a combined salad such as this Yam Yai *(literally "Big Salad").*

1 giant white radish, about 15 cm (6 inches) long
2 teaspoons salt
1 small cucumber
250 g (8 oz) cooked prawns, peeled
250 g (8 oz) cooked chicken breast or lean pork, sliced
30 g (1 oz) cellophane noodles, soaked 5 minutes in
    warm water and drained
3–4 pieces dried wood fungus, soaked 15 minutes in
    warm water and drained
100 g (3$^1/_2$ oz) beansprouts, blanched 5 seconds in boiling
    water and drained
$^1/_2$–1 cup shredded round or celery cabbage
2 hard-boiled eggs, sliced
large handful mint sprigs

Sauce:
3 tablespoons Thai or Vietnamese fish sauce
1$^1/_2$ tablespoons lemon juice
1 heaped teaspoon sugar
1 clove garlic
1 red chilli, sliced

Slice the radish thinly, sprinkle with salt and leave aside for 10 minutes. Rinse, squeeze dry and set aside. Just before serving, grate or shred the cucumber, leaving on the skin. Arrange all other salad ingredients on a platter and garnish with mint.

Combine **sauce** ingredients and blend until fine. Pour the sauce over the salad, mix well and serve immediately.

Gado Gado Salad

*Indonesians prefer salads of lightly cooked vegetables served with a tangy coconut milk sauce, or mixed with fresh coconut pounded with chilli, shallot and dried shrimp paste (*Urap*). The amount of vegetables in this* Gado Gado *depends on the number of people being served; use your own judgement, and take care not to over-cook the vegetables.*

waxy potatoes, boiled whole, skinned and sliced
round or celery cabbage, coarsely chopped and blanched
long beans or green (French) beans, cut in 4 cm (1½ in)
    lengths and blanched
water convolvulus or spinach, blanched
beansprouts, blanched 5 seconds in boiling water and drained
cucumber, skin left on and sliced diagonally
hard-boiled eggs, peeled and sliced
1–2 squares hard bean curd, deep fried until golden, diced
deep-fried prawn crackers *(krupuk)* to garnish

Sauce:
2 tablespoons oil
¼ cup very finely chopped shallots or onion
1–2 teaspoons fresh chilli paste
1 teaspoon dried shrimp paste
1½ cups fresh or tinned coconut milk
½ cup crunchy peanut butter
1–2 teaspoons palm sugar or brown sugar
1 tablespoon dried tamarind pulp soaked in ½ cup water and strained
salt to taste

Allow cooked vegetables to cool to room temperature, then arrange all salad ingredients except *krupuk* on a platter. Make **sauce** by heating oil and gently frying shallots, chilli and dried shrimp paste for 4–5 minutes. Add all other ingredients and simmer until thickened. Cool and pour over the salad. Garnish with *krupuk*.

Stir-fried Vegetables

The secret to stir-fried or braised vegetables is to cook them over very high heat to seal in the moisture and retain the maximum amount of flavour and texture. A wok is indispensible, so when you toss the vegetables they fall back in the pan and not out over the side.

500 g (1 lb) leafy green vegetables, or a mixture of vegetables
　differing in colour and texture such as baby corn, bamboo shoot,
　broccoli, capsicum, carrot, cauliflower, Chinese kale, lotus root,
　snow peas, soaked dried black mushroom, water chestnut
2 tablespoons light vegetable oil
2–4 cloves garlic, sliced
1–2 slices fresh ginger, finely chopped
$^1/_4$ cup water, light chicken stock or water from soaking
　dried mushrooms (optional)
splash of light soy sauce or 1–2 tablespoons oyster sauce
pinch of sugar

Wash the vegetables. Cut leafy greens or beans into 5 cm (2 inch) lengths. Cut the root vegetables in thin diagonal slices and break broccoli or cauliflower into small florets.

If you enjoy the taste of garlic, use 4 cloves. Heat the oil in a wok and gently fry the garlic until golden and crisp. Drain and keep aside. Fry the ginger gently for a few seconds, increase the heat and add the vegetables. Turn the vegetables constantly for 2–3 minutes until just cooked.

If you prefer to braise the vegetables, as soon as they are sealed by frying in the oil, add a little of the water, stock or mushroom liquid, reduce the heat slightly and cover the wok. Simmer until the vegetables are just cooked, stirring and adding a little more liquid as necessary to prevent the vegetables drying out. Season with soy sauce or oyster sauce and sugar. Put in a serving dish and garnish with the reserved crisp fried garlic, if desired.

Vegetables in Coconut Milk

Vegetables are simmered in coconut milk with various seasonings in most parts of Southeast Asia. The seasoning may be as simple as a few sliced shallots, chillies and a tablespoon of dried prawns or dried anchovies, or it may include freshly ground spices. This combination of a starchy vegetable, such as sweet potato, with leafy greens is a regional favourite.

375 g (12 oz) pumpkin, sweet potato, taro, pumpkin,
    young jackfruit or potato
200 g (7 oz) leafy greens (spinach, water convolvulus) or long beans
8–10 shallots or 1 large red or brown onion
2 fresh red chillies or $1/4$–$1/2$ teaspoon fresh chilli paste
$1/2$ teaspoon dried shrimp paste
$1^{1}/2$ tablespoons light vegetable oil
1 heaped tablespoon dried prawns, soaked 10 minutes and drained
$2^{1}/2$ cups fresh or tinned coconut milk
$1/2$ teaspoon salt

Peel and cube the pumpkin or substitute; if using jackfruit, prepare as directed on page 32. Wash and dry the greens or beans and cut into 5 cm (2 inch) pieces.

Pound or blend together the shallots, chillies and shrimp paste until fine. Heat the oil in a saucepan and gently cook the ground mixture for 3–4 minutes, stirring frequently. Add the dried prawns and stir fry for another couple of minutes. Put in the coconut milk and salt and gently bring to the boil, stirring constantly. Add the vegetables and simmer uncovered until cooked. Serve with rice and other dishes.

Index

Abelmoschus esculentus 46

Allium ascalonicum 45

Allium cepa 44

Allium fistulosum 44

Allium sativum 43

Allium tuberosum 42

Amaranthus gangeticus 5

Amaranthus tricolor 5

Amoracia rusticana 48

Angled Gourd 22

Apium graveolens 55

Artocarpus altilis 47

Artocarpus heterophyllus 47

Bamboo Shoot 31

Bambusa sp. 31

Banana Bud 49

Basella rubra 6

Beansprouts 37

Benincasa hispida 26

Bitter Gourd 23

Bottle Gourd 24

Brassica alboglabra 20

Brassica chinensis var. parachinensis . . 14

Brassica juncea var. rugosa 17

Brassica oleracea 20

Brassica pekinensis var. cylindrica 15

Brassica rapa subsp. chinensis 16

Capsicum .51

Capsicum annum 51

Capsicum annum cv. group longum . . .52

Capsicum frutescens 52

Carica papaya 7

Cassava (Tapioca) 29

Celery Cabbage 15

Ceylon Spinach 6

Chayote 21

Chilli 52

Chinese Celery 55

Chinese Chives 42

Chinese Kale 20

Chinese Spinach 5

Chrysanthemum 10

Chrysanthemum coronarium 10

Colocasia esculenta 11

Corn 32

Cucurbita moschata 27

Eggplant (Aubergine) 53

Eleocharis dulcis 28

English Spinach 8

Flowering Cabage 14

Gado Gado Salad 60

Garlic 43

Giant White Radish 19

Ginger 57

Glycine max 37, 40

Hibiscus esculentus 46

Horseradish Tree 48

Ipomoea aquatica 13

Ipomoea batatas 12

Jackfruit 47

Lactuca sativa 9

Lagenaria siceraria 24

Lentinus edodes 4

Lettuce 9

Long Bean 33

Lotus Root 50

Luffa acutangula 30

Luffa cylindrica 30

Manihot esculenta 29

Momordica charantia 23

Index

Moringa oleifera 48
Musa spp. 49
Mustard Cabbage 17
Nelumbo nucifera 50
Okra 46
Oyster Mushroom 54
Pachyrrhizus erosus 41
Papaya 7
Parkia speciosa 36
Phaeseolus vulgaris cv. 34
Pisum sativum var. saccaratum . . .38, 39
Pleurotus cystidiosus 54
Pleurotus fiabellatus 54
Psophocarpus tetragonolobus 35
Pumpkin 27
Raphanus sativus 19
Red-streaked Bean 34
Rorippa nasturtium-aquaticum 18
Sauropus androngynus 30
Sayur Manis 30
Sechium edule 18
Shallot 45
Shiitake Mushroom 4
Snake Gourd 25
Snow/Sugar Pea, Pea Shoot 38, 39
Solanum melongena 53
Solanum torvum 53
Soybean 40
Spinacea oleracea 8
Spring Onion 44
Stir-fried Vegetables 61
Straw Mushroom 56
Sweet Potato 12
Taro 11
Trichosantes cucumerina var. anguina .25

Twisted Cluster Bean 36
Vegetable Soup 58
Vegetables in Coconut Milk 62
Vigna radiata 37
Vigna sesquipedalis 33
Volvariella volvacea 56
Water Chestnut 28
Water Convolvulus 13
Watercress 18
White Cabbage 16
Winged Bean 35
Winter Gourd 26
Yam Bean 41
Yam Yai Salad 59
Zea mays var. rugosa 32
Zingiber officinale 57